MW00953972

How to Find Objects in the Night Sky

Allan Hall

Acknowledgements:

As always I have to thank my wife, Sue Ann, because without her prodding and help, my books would either never get done or would be far less than they are.

I would also like to thank my readers. Their emails, letters and reviews let me know that my work is appreciated which drives my desire to create more and better books.

Table of Contents

1: Introduction

One of the most difficult things for newcomers to astronomy and astrophotography is finding objects in the sky. We can pretty much all find the sun and moon, but everything else is simply different degrees of frustration.

It doesn't have to be!

In this book I will show you the three primary ways of finding things in the night sky: star hopping, altitude azimuth, and right ascension & declination.

Star hopping is exactly what it sounds like, finding a known object and "hopping" to the next one. There is a little more to it than that but that is pretty much how it works. No, you do not have to memorize a lot of objects, or really any objects to start using this method.

Altitude azimuth is really easy to wrap your head around, each object is X number of degrees up from north, and X number of degrees up from the side. Unfortunately that changes every minute but is still easy to use.

Right ascension and declination is the "hard way", but even then there are some tricks to make it much easier than you might have thought.

Once you finish this book you should be able to find any object in the night sky that you can find a star chart for, and that is all of them.

So let's get started!

How to find objects in the night sky

Section 2: The basics

2.1: Movement of the earth and sky

Our planet is a member of a solar system, a lively collection of planets and asteroids orbiting a star we call the sun. That star is one of billions of stars making up the Milky Way galaxy we sit in, out on one of its spiral arms. The Milky Way galaxy is a member of the local group of galaxies which is comprised of well over fifty galaxies, many very similar to the Milky Way. The local group is a small part of the billions of galaxies in the universe, each with billions of stars and many of those with planets orbiting them.

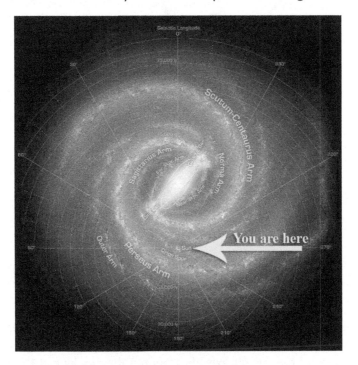

Add these objects to the planets in our own solar system and the moons of those planets and we have a lot of objects we can observe.

The earth rotates around an axis. This axis is like a line going through the earth from the north to south poles and extending out into space. The point in space seems to be at different places depending on where you are on Earth.

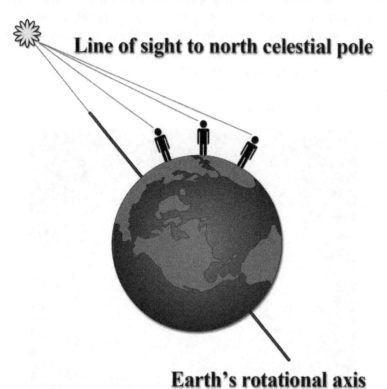

Line of sight to north celestial pole

Earth's rotational axis

In the previous figure the star in the upper left corner represents the North Star, Polaris, the star that is almost exactly in line with the center of the Earth's axis of rotation. When looking at the night sky (or in the day for that matter) the sky seems to rotate around this point in space in the northern hemisphere (from the equator north to the North Pole). There is a similar point in the southern hemisphere (from the equator south to the South Pole).

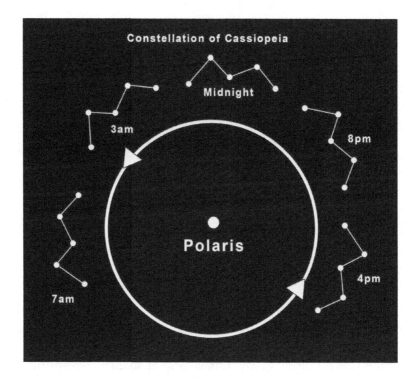

In the above figure note that the constellation Cassiopeia not only moves across the sky, but rotates as it does so. This motion is centered in the northern hemisphere on the North Star, Polaris. The center of the "W" always points towards Polaris.

What this means for you is that as an object you may want to view is moving across the sky, it is also rotating around the rotational axis. This is important to remember as you try to match what you see to maps and images of objects.

This rotation completes one full turn approximately every 23 hours, 56 minutes, and 4 seconds and is referred to as one sidereal day.

While the Earth is rotating, it is also orbiting the sun.

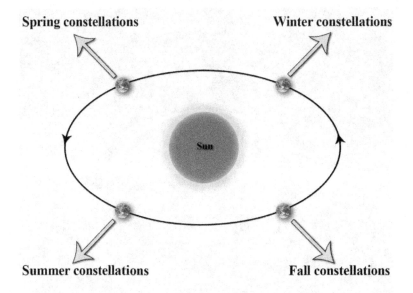

Spring constellations

Winter constellations

Sun

Summer constellations

Fall constellations

It takes a year for the Earth to completely orbit the sun. At different times of the year the night side of the Earth is facing a different direction in space. This is why at midnight every three months you can look directly overhead and see a different constellation. This is also how our ancestors used the sky as a calendar to know when to plant their crops, among other things.

Since the east/west side of the Earth faces the sun this also explains why the stars very close to the north and south celestial poles (such as Polaris and Sigma Octantis respectively) remain up all year long. Constellations such as Cassiopeia may circle the celestial pole but depending on your latitude they may never dip below the horizon.

How to find objects in the night sky

All objects in the sky are found one way or the other by using degrees. There is an easy way to estimate the number of degrees between objects in the sky with no tools. At least no tools you were not born with.

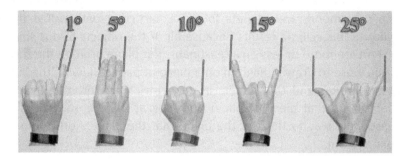

Hold your hand up at arm's length. Your pinky appears roughly 1° wide, the width of three fingers is 5°, your fist is 10°, the distance between your index and pinky tips when extended is 15° and the distance between your thumb and pinky tips appears approximately 25°.

Larger distances are easy as well since the entire sky around the earth would be 360°. This means that the part of the sky you can see is roughly 180° from horizon to horizon, straight up in the air would be about 90° from the horizon, and half way between directly overhead and the horizon is about 45°.

If you have ever seen an astronomer with his fist raised in the air he could have been either cursing the clouds or measuring the sky. ☺

While this method is not the most accurate in the world, it will allow you to quickly and easily navigate anywhere in the sky using virtually any star map available.

2.2: Movement of the sun, moon, and planets

As we discussed, all objects in space appear to rotate around us. The sun, moon, and planets follow a particular path called the ecliptic. This ecliptic results because all of the objects in our solar system are more or less on a reasonably flat plane orbiting the sun. If you were to draw a diagram on a piece of paper showing the sun at the center and the planets in concentric rings orbiting it and then lift up that piece of paper and look at it edge on, that is a reasonable way to think of the plane that the planets orbit in our solar system.

As you look at objects in the sky you can see this because the planets all seem to chase each other in the night sky. In addition the moon will follow (or proceed) the sun in its path across the sky.

Computerized telescopes that track objects in the sky will have several different speeds so they can track different objects. Even the most basic tracking telescope will have both a lunar rate (for tracking the moon) and sidereal rate (for tracking the stars and other objects).

If you wanted to track faster objects such as satellites or the international space station you may be out of luck. Most consumer mounts cannot track at this speed. Even the very few that can require some "tweaking" to get them to do it.

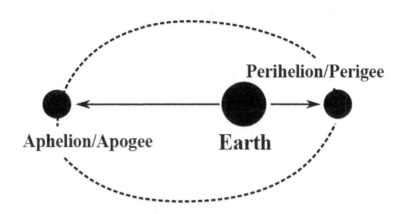

When discussing orbital paths there are a few terms you will hear frequently. These terms vary depending on what the objects are orbiting. When orbiting the Sun, an object that is at its closest orbital distance is said to be in perihelion. When it is at its farthest orbital distance, the term is aphelion. When the object is orbiting the Earth instead of the Sun these terms change to perigee and apogee respectively.

To make things even more interesting, when orbiting the moon during the Apollo program those same terms became pericynthion and apocynthion.

Some planets such as Mars approach the Earth at vastly different distances. Depending on where each planet is on its orbit around the sun, the distance differences can make observing very difficult in that you can see virtually no surface details. Other times Mars will be close enough that surface details are readily apparent even in medium sized telescopes.

Some planets such as Mercury will appear as little more than a speck on the best of days.

You can see that if you decide to observe the planets, you will need to find out where they are in their orbital path to maximize your ability to coax detail out of them.

2.3: Light pollution and your ability to see

Light pollution is caused by street lights, signs, and your neighbor's porch lights bouncing back up into the atmosphere. This light reflects off of dust and water vapor in the air and makes it harder to see astronomical objects.

In late March one year we had a storm that took out the power for many miles. I used this opportunity to take two pictures. Both images have the same exposure and were taken at the same time at night. The difference is that one was taken on March 28th 2014 at 10:45pm after the power was knocked out. The other was taken on March 29th 2014 at about the same time when the power was back on. The difference was staggering.

Inspecting the images I could see first-hand that not only was the sky brighter and the stars harder to see, but the trees were brighter too! This I attributed to the light from stores, houses and streetlamps bouncing off the pavement up into the trees on its way into the sky. It was difficult to imagine there was this dramatic a change until I actually saw it myself.

These effects are why people drive many miles away from the cities in order to view celestial objects. The further you can get from any lights, the better. It can also help to get hills and even mountain ranges between you and the lights. If you are stuck in town try to get into a park with some tall trees on all sides to block the city lights. Failing this, try a safe rooftop such as on the top level of a parking garage to see if you can get above as much of the light as possible.

Not all of us can travel too far however and even if we can, there is still some light pollution. Fortunately there is something that can help, light pollution filters.

Light pollution filters are designed to block the specific wavelengths of light generated by streetlights, signs, etc., while allowing light from the celestial objects to pass through unharmed. This improves our signal (light from celestial objects we want to see) to noise (light pollution) ratio making the objects stand out better against the night sky.

To see light pollution, you merely have to look up at the sky towards a large light source. An example of light pollution would be the glow in the sky surrounding a football field at night. This light can be minimal out hundreds of miles from a town, or it can be horrible in the middle of a large city.

Taking two pictures of the same deep space object, shot on the same night, from the same location, minutes apart, with the same telescope and camera without and then with a light pollution filter can clearly show the problem.

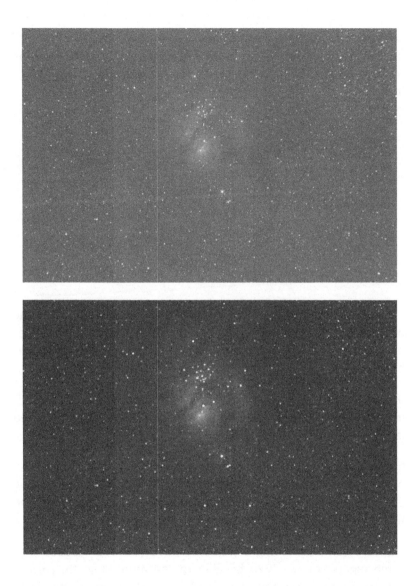

In the two pictures we took, the one taken with a light pollution filter makes the nebula stand out from the background substantially better than in the one taken without.

Unfortunately good LP filters are not cheap, but they are worth every penny. Can you view without them? Absolutely! But I highly recommend one as it can substantially improve your viewing in all but the darkest skies. While no filter will eliminate all light pollution from your view, having a quality LP filter certainly will help reduce it.

I cannot stress enough that if you live anywhere except out in the middle of the desert or high up on a mountain, hundreds of miles from the nearest department store, they can provide a serious benefit.

If your adapter will allow the use of a 2" filter you should use it instead of the smaller 1.25" filter to prevent vignetting (discussed later).

When you go to purchase a light pollution filter, be aware that they make them both for visual and imaging. While filters for imaging use may help when viewing, my experience has been that the filters for imaging are a little too strong for visual use.

Now we come to seeing conditions. Have you ever noticed that some nights you seem to be able to see more stars than other nights? Some nights the stars really seem to twinkle and other times they are just still pinpoints of light? This is probably due to the seeing conditions.

Many things affect the seeing conditions such as the amount of water vapor in the air, not just down here where we are, but at one hundred thousand feet, and everywhere in between. Even the temperature at different altitudes can affect the seeing. This is why most large telescopes are built at the top of mountains or in the desert.

Now you don't have to travel to get good conditions, you just have to watch the weather. Among other things, astronomers use what is called a Clear Sky Chart (CSC) to see what the conditions are likely to be for the night. You can see what one looks like by visiting the following website and searching for an observing site near you:

http://cleardarksky.com/csk/

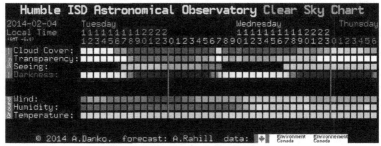

Figure 1: A sample clear sky chart.

The chart is read by the hour which you can see above under the words Tuesday, Wednesday and Thursday, in 24 hour time. Each square is colored and for the top six squares from top to bottom the darker blue the better, the lighter the worse. So if you had all six of eight squares for a specific time in the darkest blue possible, that would be an estimated perfect night for astronomy.

So let's talk about what each box means from top to bottom. Cloud cover is the estimated amount of cloud cover for that particular time, white being completely overcast, light blue such as 16 under Tuesday in our example chart being 50% cloud cover and the darkest blue being seeing no clouds anywhere. That was the easy one. ☺

The next box down is Transparency and that is pretty much a measurement of the amount of water vapor in the air. Yes, clouds can contain very high amounts of water vapor, but no clouds in the sky does not mean a lack of water vapor, or even a reduction of water vapor.

Next is Seeing and that is a measurement of the air turbulence caused by different thermal layers in the atmosphere. You may have noticed that as you look down a long stretch of road in the summer you can see that things in the distance are distorted by the heat rising from the road. This demonstrates how light can be refracted by the different thermal layers which can make it very difficult to image planets and stars. What happens is the rippling effect of looking though the thermal layers tends to make the stars twinkle.

Darkness is the next item down and it is just that, a box that tells you how dark it will be relative to daytime. There are basically two things that affect this, the sun and the moon. If neither the sun nor moon are anywhere near rising, this will be very dark blue. If the sun has long since set but the moon is up, then it will be lighter blue. If the sun is up, it will be white.

After darkness we have Wind. The more wind, the harder it is to get a good view as the telescope will be buffeted by the wind. Larger telescopes also have more of a problem than smaller telescopes with wind.

The next two, Humidity and Temperature are not as important really as there is not much you can do about either. Temperature will be directly related to where in the world you are and the time of the year so normally watching it from one night to the next will make no difference. Humidity doesn't matter much as that is a measurement of water vapor at ground level and all we care about is total water vapor level shown in transparency, we don't really care whether it is at ground level or fifty thousand feet.

Please keep in mind that these are forecasts so take them with a grain of salt and be prepared for the unexpected. I always tend to carry more cold weather gear than I expect to need and also some full size trash bags to throw over things should rain come from nowhere.

There are widgets (put the CSC on your Windows desktop), apps (iOS and Android) and programs (Weather Ninja by CCDWare) to display the Clear Sky Chart along with, of course, its web page.

Remember that programs such as these are merely forecasts and as such can be very wrong at times. It is advisable for you to carry trash bags or tarps to cover your observing gear should the weather turn bad quickly. I also tend to carry a plastic bag or two for things such as my cell phone and tablet.

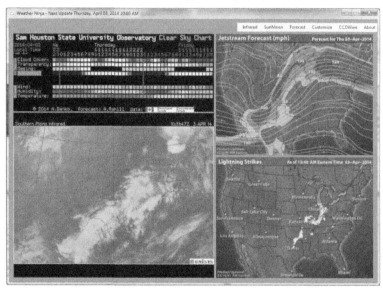

Figure 2: Weather Ninja main screen.

Another useful tool that I use in conjunction with the clear sky chart is satellite imagery of the area I am in. Weather Ninja is a piece of software that has this feature built in as do many weather websites. The idea is that the satellites take images showing the clouds over the area every few minutes. These images are then displayed in succession to give the appearance of motion. This allows you to not only see where the clouds came from, but to estimate where they are going.

How to find objects in the night sky

Section 3: Navigation - Finding your targets

How to find objects in the night sky

3.1: Star Hopping

Star hopping is simply finding an object you know, then using it to navigate to another object in the sky.

What if you did not know which finger on your hand is called your ring finger? What if I then told you it was between your pinky and middle finger? That is using "finger hopping"!

Star hopping requires a star map that shows two things, a target you want as your destination, and some other objects you either know or can find in the sky. If you have both of those things then you are set. The star map can be a printed piece of paper, or something on a computer/tablet.

So what if you don't know any names of the stars? We can do all of this without any star names at all. Let's give it a try.

To do this we are going to use two pieces of software; Stellarium and C2A. Stellarium will represent our night sky, what we are actually seeing when we go outside and look up. The C2A chart will of course be our chart that we will use to help us navigate.

In this first example we will be looking for Messier 13, the Great Globular Cluster in Hercules.

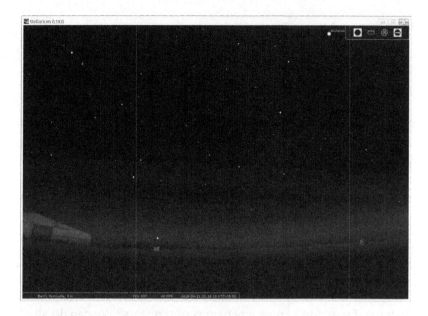

In the above image you can certainly make out the ground, a building over on the left, and some stars, but what stars are they and where do we start? Let's take a look at the star chart here:

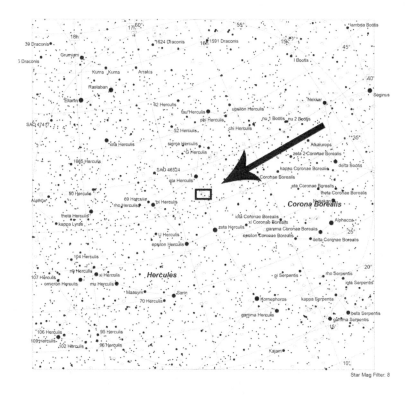

Wow! That sure is a lot of stars and stuff, how the heck are we going to figure any of this out?

Start by looking at the rectangle where the arrow is pointing. I added the arrow and made the rectangle a lot bolder than it normally is so it would be easy to see in this small book.

Now let's break things down to make them easier, what are some of the brightest stars shown on the star chart close to the target we are looking for?

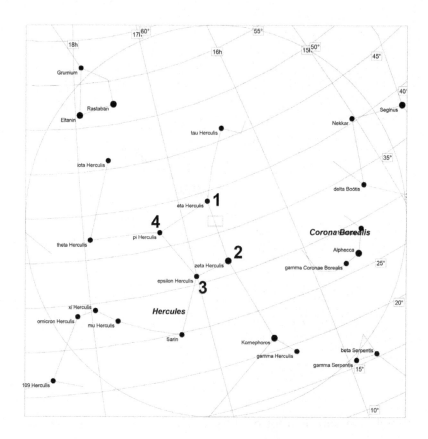

In this image I removed all the objects except for the brightest stars so we could concentrate on just what we needed. I also put the numbers 1, 2, 3 and 4 next to the four brightest stars around the object we were looking for. You don't have to reprint a map like I did, just mentally use the easiest objects for your search.

In this case, the object is right on a line that makes a rectangle in the sky. Since these are some of the brightest stars in the sky, and they make up the center of the constellation Hercules, they should be pretty easy to find.

Let's see if we can find those four stars in the sky...

The first problem we have is that it is an awfully large sky with an awfully lot of stars. We need a way to narrow down where to look or this may take all night.

Star charts are almost always printed with north being up, so that gives us a rough direction of where this square will be facing.

I can use a Planisphere and dial in the date to see that Hercules will be near the north east horizon around 10pm.

Without a planisphere (which you really should have, something like the Miller Planisphere for your latitude will serve you well for years) you could look at your C2A or other charting program and see roughly the same thing. This includes apps on your phone or tablet such as Star Walk or Sky Safari.

Lastly, you could go online to in-the-sky.org and enter your location and the object or constellation you are looking for to see where it will be at a specific time. In this shot, I told it to use my internet browser's location, for 10pm tonight then scrolled around until I found the constellation Hercules as shown below.

This shows us that Hercules will be just above the horizon to the north east at 10pm.

Once you get used to this and have done it a few times you will spend more time looking at the sky and less time trying to figure out where your reference points are.

Now that we have a rough idea of where to look in the sky, we can start looking for our pattern.

28

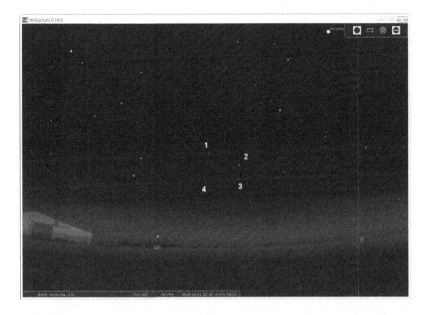

Looking at the sky we find our four stars, and I know they are the right ones because they are not only in the same pattern as the star chart shows, in the place/time our online search shows, but also if you look at the size of the stars in the star chart you will notice that #2 is the brightest (biggest on the chart), and indeed the brightest star in the sky is in the place it should be in the pattern.

Now that we know where we are in the sky, where exactly is our target?

Looking at our star chart it looks like the target is about one third of the way between #1 and #2 stars, which would put it right about....

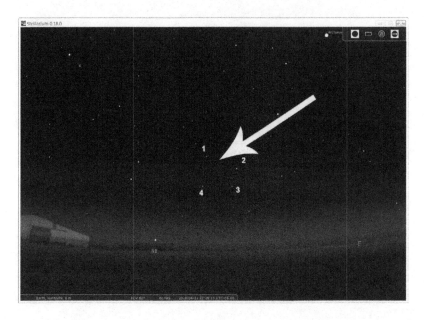

here. So if we zoom in to that area and look, sure enough, there it is.

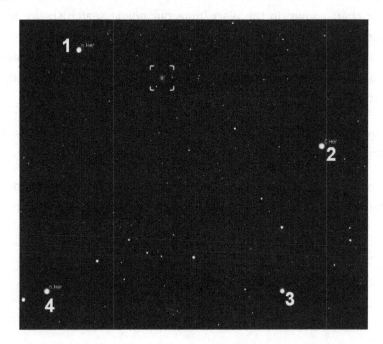

That was pretty easy, wasn't it? There are however things that can make it more difficult, some of which you might not expect. What happens if you get out under a really dark sky? Is it possible that the seeing is too good? You bet! Let's take a look at that scenario.

Notice in the image above we have a LOT more stars showing and that does make it harder to find the stars we want. Fortunately, with a little careful looking we see the same four bright stars in the same place.

In this case it is just a matter of paying more attention to the brightest stars while mentally ignoring the dimmer ones.

Unfortunately the opposite is not true. If the seeing is so poor you cannot find any reference objects then you are out of luck. If this is the case then you are probably in town where there is a lot of light pollution or under clouds. You can't do anything about the clouds but if it is the light pollution just drive a few miles out of the city and you should be good.

How to find objects in the night sky

So now that we have found our target, all we have to do is use a wide enough field-of-view eyepiece in our telescope (high enough number such as a 25mm) to start with so we can get at least stars #1 and #2 in the view at the same time.

Now we move the telescope until we think the target location is in the center of the field-of-view and slowly increase the magnification (use an eyepiece with a lower number such as 15mm).

If you slowly step through your eyepiece magnifications centering the target each time, and assuming that you have dark enough skies and a large enough telescope, you should be able to find any object in the night sky with this technique.

Let's review...

1) Find a pattern on the star chart of bright stars around or near our object, mentally figure out how we will "hop" from one object to the next to find our target.
2) Find out when and where our pointer objects (the central box in Hercules in this case) will be in the sky using a planisphere, app, or online.
3) Find our pointer objects in the sky.
4) Mentally draw a line from star #1 to star #2 in the sky.
5) Estimate the distance between #1 and the target on that line.
6) Start with a very low power eyepiece and step through our navigation in the sky just like we did on the star map.

That one was pretty easy because it was easy to see, even with the naked eyes under clear skies, and since it was right on a line between four of the brightest stars in the sky.

How to find objects in the night sky

Let's try something a bit harder, M101.

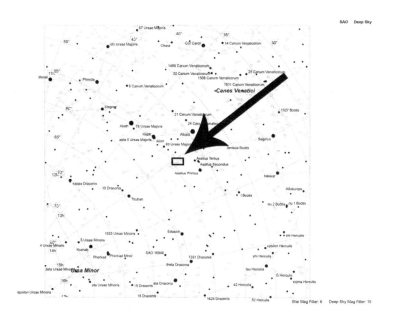

Above is our standard star chart with the object highlighted.

There are some interesting things in this chart. Let me simplify the chart and make a couple of observations for you.

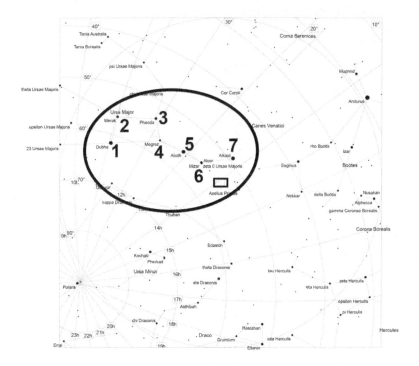

In this version of the same chart, the oval is circling the "Big Dipper" or more properly Ursa Major, the Great Bear. Also note all the lines that meet at a point in the lower left, this is Polaris or the North Star. You should notice that the constellation is almost right above the Polaris with two stars (#1 and #2) on the far left of the constellation pointing almost right at Polaris.

Looking at the image above, can you see it? Do you see the upside down ladle with the edge of the cup pointing down towards another star that is very close to the north marker?

Spend some time and don't look at the next page until you either see it, or give up.

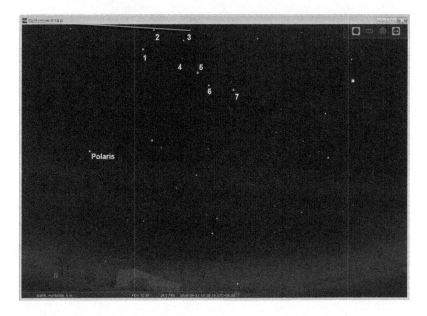

If you didn't see it, flip back and forth between this image and the previous one and see what you missed.

Now that we have this reference we need to navigate to our target. To start, we will take a close-up look at our last star chart.

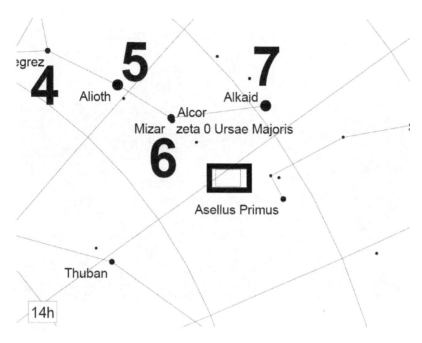

The rectangle in the center is our target. Note that it makes almost an equilateral triangle (triangle where all sides are the same length) with stars #6 and #7.

Zooming in a little in the sky (by using a higher power or lower number eyepiece) we see something similar to the following:

It is pretty easy to match up #6 and #7 stars, Asellus Primus is pretty easy too since you can clearly see the little triangle of stars which includes Asellus Primus in both the sky and the star chart.

The white arrow is where our target is, even though you probably can't see it in the image above, and maybe not even in the sky at this magnification, but it is there.

Notice that if you draw a mental line from star #6 to Asellus Primus, our target will be just to the left of that line, just a little bit short of half way between Asellus Primus heading towards star #6. This position also makes a pretty convincing equilateral triangle between our target and stars #6 and #7.

So again we have found our target without knowing hardly anything about the night sky. Easy!

As we have done both times before, pay attention to the brightest stars that will get you close to where you need to be, and try to use geometric shapes not only to finally find your target, but to find your references along the way.

All star hopping is made up of those types of procedures although some people do things a little differently depending on their knowledge of the sky, how they were taught, and what works best for them.

The important thing is not following some specific procedure, but that you see and understand the process so that you can adapt it to what works best for you.

As you learn more of the sky, the stars, constellations, etc; you will be able to star hop faster and easier. The more references you try to use, the more you will remember. I will bet you that you know the internal shape of the Hercules constellation and how many stars are in it, don't you? ☺

3.2: Altitude Azimuth

Altitude & azimuth is a very simple system that measures the angle in degrees of an object above the horizon (altitude) and the angle in degrees of an object from north in a clockwise direction (azimuth).

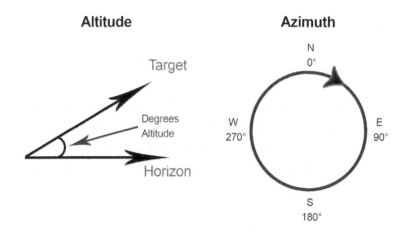

An example would be the North Star, Polaris. At my location Polaris would have an altitude of approximately 30.5 degrees and an azimuth of approximately 0 degrees since it is almost exactly north.

Altitude and azimuth make it very fast and easy to find any object in the sky, but it does have one big drawback and that is that these coordinates are always only valid for a given date, time, and location since the earth is constantly rotating (with the exception of Polaris of course). This means that if an object is at an altitude of 30 degrees with an azimuth of 94 degrees, in an hour that will have changed to different coordinates as the object will have changed position in the sky.

Think of it this way: early in the morning the sun rises in the east at 90 degrees (not really, but for the purposes of this example let's say it does). Further, let's say a day is exactly 8 hours long at this time of year. That means the sun moves at 180 degrees in 8 hours or 22.5 degrees per hour. So if the sun in our example was at altitude 0 at 8am, it would have an altitude of 22.5 degrees at 9am, 45 degrees at 10am, 67.5 degrees at 11am and 90 degrees at noon, directly overhead. Yes, that would be a really short day, but it should help you visualize what is happening with this coordinate system.

So you may ask, since these change every second, why would I want to use them for anything? Because they are simple and most astronomy programs (and apps) will show you the altitude and azimuth in real time, making them easy to find fast, even for beginners.

Take for example our last star hopping example, M13. When you do a search in C2A for that object you get a window that looks like this:

Note on the right side the Az (azimuth) and Alt (altitude). So to find the general area of this object all I have to do is look approximately 63 degrees clockwise from north, and then approximately 31 degrees up from the horizon and there it is.

Sure, I can't get too exact, but if I look a little less than 45 degrees (which is pretty easy to guess) between north and east, and then a little over 45 degrees up towards right over my head, that will get me close enough to find my reference points.

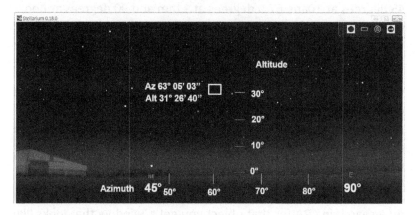

Looking at our night sky again, NE is 45° and E is 90° so it is pretty easy to mentally see where 60° would be. Then we can use our outstretched clenched fist like shown previously in the book to see 10, then 20, then 30 degrees up from the horizon.

This puts us really close to where our target is without breaking a sweat!

If you have an Alt/Az telescope mount (a type of telescope mount that uses only altitude and azimuth to guide you) then you could use the scales on the mount to get you right on target.

3.3: Right Ascension & Declination

Next comes the Right Ascension and Declination method which is a little harder to use, but does not depend on the date, time or location at all.

Declination is basically the same thing as latitude, or a measurement from the equator which is 0 degrees, to the poles; positive 90 degrees to the North Pole, and negative 90 degrees to the South Pole. Think of this as being projected from the surface of the planet out into space, so an object that is in a direct line above the North Pole would be at +90 degrees declination regardless of where on earth you were standing. Declination is measured in degrees, minutes and seconds from the celestial equator.

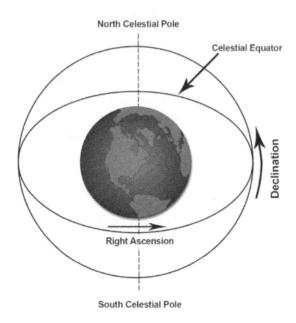

Right ascension can be a tough one for people to understand. There is an imaginary point in space called the Vernal Equinox which is the starting point of measurements in a circle running along the celestial equator (if you took the equator on earth and projected it into space, this would mark the celestial equator). If you started at this imaginary point and spun the earth one full revolution back to that same point, which would be 24 hours of rotation, always measured at the celestial equator. Right ascension is measured in hours, minutes and seconds from the Vernal Equinox towards the east.

Every object has a single RA/DEC coordinate which remains the same regardless of date, time or location. Messier 42, the great Orion nebula for example is RA 5h 35.4m and DEC -5 degrees 27'. These coordinates are the same for this object whether you are in the northern hemisphere or the southern, on daylight savings time or not, or to the east or west of the international date line.

How in the world would you be able to use RA/DEC? Most equatorial mounted telescopes have both RA and Dec setting circles built into the mount. The easy method is to point the telescope at a target you know, set the setting circles for the correct values for that object, and then move the telescope until the setting circles show you are pointed at the new target you want to find.

Just like with star hopping we need a reference point and where we want to go.

Where we want to go is easy, we just look at the same object info window we used for Alt Az in C2A and it shows us our RA and DEC right to the left of our Alt Az.

Note that in C2A it calls DEC just DE. Now we need some known points to start with, so let's use one we are already familiar with.

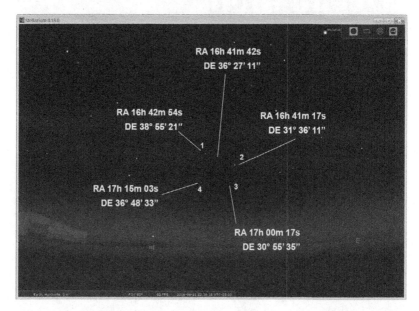

In the image above we are looking back at our night sky, and back at the Hercules constellation's central rectangle where we were looking for M13.

I have noted the RA and DE (short for DEC as we saw in C2A) for the four stars in the rectangle we had already numbered 1-4, as well as the target M13.

Just looking at this you should be able to see that as you look further away from north, the DE numbers go down, in fact when you look past east, those numbers will go negative.

So with DE, you start at 0° on the east (or west) and go to +90° at the North Pole, and -90° at the South Pole. OK, that doesn't sound so hard now that you see it in action, does it?

Just remember that the degrees start at +90° at the North Pole and decrement as you go south. I am saying that a little differently because we said earlier that it goes from +90° to the north, then it gets to 0° not only when you hit east or west looking right or left, but also when you get directly overhead looking up.

Taking a look at the RA on our image we note that as we go down towards the horizon to the east, the number increases in hours. Star #1 is at 16h 42m whereas below it star #4 is at 17h 15m.

This means that if you start on the horizon due east and go towards the horizon due west over your head, you will go backwards in RA time about 12 hours.

With just a little math you can see that if objects on the eastern horizon are RA 17h then objects on the western horizon are RA 5h.

So let's give this a try.

The first thing we do of course is polar align our equatorial mount telescope. You should already know how to do that, if not check the instructions that came with your telescope mount, or you could purchase my book *Getting Started: Using an Equatorial Telescope Mount* which will cover everything you need to set up the mount.

Now we get the telescope pointed to our known reference point which in this instance will be our #1 star in the image on the previous page.

Once the star is in the center of our viewfinder, we can set the setting circles on our telescope mount to match the known coordinates of this target. First we do the Right Ascension which is 16h 42m.

Since we are visually trying to find a target we do not need to be overly exact so since 45 minutes is three quarters of an hour and our current reference point is 42 minutes, I set the circle to approximately three quarters of the way between 16 hours and 17 hours.

Now I move to do the Declination which is about 38° 55′ or almost to 39°.

The next trick is to use our slow motion controls (the little knobs that move the telescope on an all manual mount, or the arrows on our hand controller for electronic models) just a little bit. We want to make sure that the telescope moves and the pointer moves over the scale.

Sometimes there is a lock that makes the pointer and scale move together and we want to make sure that does not happen. We need them to move independently.

Once that is checked and they do move independently, simply move the telescope until the RA shows the position on the scale we want which is 16h and 41m, about one minute away, so realistically we can just leave it where it is, but double check it.

If we were instead looking for star #4 we would need to move the telescope until the dial showed 17h 15m, or one quarter of the way between 17h and 18h.

Next we move the Declination of the telescope until our scale in that direction shows us 36° 27', or about half way between 36° and 37°.

Lastly, we should be able to look through our eyepiece and see our target!

On many computerized mounts you can also enter the coordinates directly to navigate to an object that may not be in the built in catalog of objects.

If you are not using an EQ mount you can still use this system the same way, by finding a known target and moving from there. Even Alt/Az mounts that are computerized will normally allow you to enter coordinates in RA/DEC to navigate to objects.

3.4 Recapping

Reviewing the highlights:

1) Star hopping always works and has no coordinates. This is how most visual astronomers work.

2) Alt Az coordinates change throughout the night, M13 may be at Alt 31° and Az 63° at 10pm, but at 11pm that changes to Alt 43° and Az 66°. This is a helpful tool for visual astronomers to find objects fast.

3) RA and DEC are always the same. M13 is RA 14h 41m 42s and DEC 36° 27' 11" now, tomorrow, at 8am and at 10pm, and forever. This is the method used by most astrophotographers, and most professional astronomers. These people typically just put the coordinates into a computer which points the telescope for them.

One point I want to make sure you get before we move on is that you do not have to be exact here. Do not worry about the difference between 16h 41m and 16h 42m, your telescope is probably nowhere near accurate enough to worry about it.

Even if you were using a very expensive scientific grade telescope that could absolutely be accurate down to tenths of seconds, YOU won't need it.

If you are using the standard 25mm Plossl eyepiece with a 52° field of view, you will be seeing about 1.3° of sky in your eyepiece. That works out to about 5m 12s of sky, so being off a minute or two will make absolutely no difference in getting your target in your eyepiece.

Section 4: Seeing your target

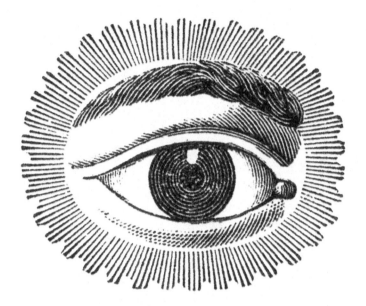

4.1: Finding is only half the battle

Once you navigate to your target you still have to see it. With targets such as the M13 we have been working with, that is easy as long as you are not legally blind or observing from downtown Houston.

With other targets it can be a little more difficult. M101 we also talked about is much harder to see than M13 but still something that most people can make out with minimal equipment and a semi-dark viewing location.

Others are a lot harder and the absolute best cure for this (aside from spending obscene amounts of money on new gear) is experience. Very experienced observers can see things that beginners have no hope of seeing, even with superior equipment.

Let me clarify that previous statement: a very experienced observer can see things with a $300 telescope that a beginner cannot see with a $1000 telescope. Fortunately we can all gain experience so just be patient.

There will always be targets you just cannot see with your equipment and at your location. This is why so many of us get caught up in spending large sums of money on bigger and better equipment, and on traveling chasing darker skies.

There are however some tricks I can share with you to help you get started.

4.2: Dark Vision: Adapting to the dark

Your eyes are composed of two types of organs for seeing; rods and cones. The rods are much more sensitive to light and you have many more of them but they are not sensitive to color. The cones are sensitive to colors but are far less sensitive to light and far fewer in number. Since the rods cannot see red light very well as it is a color, you can safely use **low levels** of red light without harming your night vision. One word of caution; green is also a color but the eye's receptors are far more sensitive to green than to red or blue so stay away from any green light sources.

It takes approximately 30 minutes for your eyes to fully adapt to the dark. My experience has been that the more you observe the faster and better your eyes will adapt.

I have been out when there was an astronomy class outside at the observatory and my tired old eyes could see my surroundings just fine while the college kids were running into things blind as a bat. Give them an hour, they could see just as well as I could with only ten minutes of adaptation.

Your eyes can also recover from light exposure faster as you get more accustomed to the dark. While I certainly can lose my night vision if someone shines a car's headlight at me, I will recover and be able to see in the dark faster than someone who never spends time in the dark.

Some tricks to remain dark adapted are to use only red lights. This may seem obvious until you understand what I mean. What happens when you are outside and you need to go inside to use the restroom? Buy a few nightlights that use small bulbs like the ones on old Christmas tree lights, get some red bulbs and put them in the nightlights. Place one nightlight in the restroom, and one in each room or hall leading from the outside door to the restroom.

You can also put some form of red plastic over a white nightlight to get the same effect. Just be sure that the light does not get hot enough to melt the plastic or worse, set it on fire.

Whatever you do, stay away from the refrigerator, microwave, and oven as they all have white lights that come on when you open them. It is not all white lights though; your TV can destroy your night vision just as easily.

This is also true for cell phones and tablets. It is best to turn the brightness down all the way before dark so if you do turn them on during your observing session the impact is minimal.

Astrophotographers are weird as some of them will prefer to use very dim regular lighting instead of red lights.

A good example would be that my laptops are covered with ND (neutral density) filters instead of the typical red filters called Rubylith. This is because when the screen is tinted red, it is very difficult to see faint nebula details in an image. ND filters solve that problem.

4.3: Averted vision and tapping

Averted vision is a technique used to catch a glimpse of an object that is extremely faint. To use averted vision, look at where an object should be in your eyepiece, and then look away over towards one edge of the eyepiece to try to get a glimpse of the object with your peripheral vision. This can sometimes pull out details that you could not directly see. When first using this technique, you may need to try several times to see the effect.

It is interesting to note that we think the first report of using averted vision was from Aristotle in his book *Meteorologica* at around 325BC.

Another interesting technique is tapping. This involves looking at an object while you tap on the eyepiece enough to make things move just slightly. This can make very dim objects jump out that you could not otherwise see since our eyes are more sensitive to objects in motion. With this technique we are tapping very lightly with one finger on the eyepiece. Start out barely touching the eyepiece and slowly increase the force you use until the stars in the viewfinder move just enough to make them blur. If you tap too hard you may knock the telescope off target completely.

4.4: Field of view

An important concept that some people take for granted is the field of view. This is thought of by beginners as the magnification shown in the telescope.

It really is not that simple however because magnification and field of view are related but not the same.

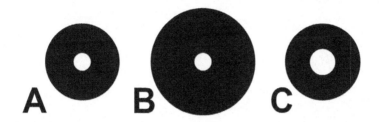

In the above image there are three views of the same target. The first is our base view and let's say it is a 15mm eyepiece with a 50° field of view. The magnification is 50x.

In B, the magnification is the same, but the field of view is larger. Let's say this is a 15mm eyepiece but with a 68° field of view.

With C we have the same 50° field of view as in A but the target appears larger. We could say this is a 12mm eyepiece.

So why is this important? Because when you are trying to find an object you want to start out with an eyepiece with a wide field of view and low magnification, then work your way down.

Many beginners will start out with their higher number eyepiece such as a 25mm and then move down to a smaller one as they navigate closer to the target, and that is absolutely the right idea. It does not take into consideration the field of view directly, it assumes an equal field of view for each eyepiece.

If you are only using the eyepieces that came with your telescope then that may be a true assumption, or maybe not. If however you bought additional eyepieces then it probably is not true.

It is also important to determine what eyepieces you want to purchase in the future.

5: More information

How to find objects in the night sky

5.1: Where to go from here

Oh boy, are there a lot of places you acan go so here are some suggestions:

Astronomy equipment:

Orion Telescopes	-www.telescope.com-1-800-447-1001
Agena Astro	-www.agenaastro.com-1-562-215-4473
Oceanside Telescope	-www.optcorp.com-1-800-483-6287
Shoestring Astronomy	-www.shoestringastronomy.com
Astromart (used)	-www.astromart.com
ScopeStuff	-www.scopestuff.com

Online forums:

Astronomy Magazine	-www.astronomy.com
Stargazers Lounge	-www.stargazerslounge.com
Cloudy Nights	-www.cloudynights.com
Astromart	-www.astromart.com/forums/
Telescope Junkies	-www.telescopejunkies.com

Specializations:

Spectroscopy	-www.rspec-astro.com
Radio Astronomy	-www.radio-astronomy.com
Photometry	-www.citizensky.org
Astrophotography	-www.allans-stuff.com

Planetarium software:

TheSkyX	-www.bisque.com
Starry Night	-www.starrynight.com
Stellarium	-www.stellarium.org
Cartes du Ciel	-www.ap-i.net
C2A	-www.astrosurf.com

Session planning software:

Astroplanner	-www.astroplanner.net
Skytools	-www.skyhound.com
Deep Sky Planner	-www.knightware.biz

5.2: Index

5.3: Glossary

A/D converter (ADC) - Analog to digital converter. A camera sensor records light as an analog signal which the A/D converter then converts into digital information.

Achromat – A type of refractor typically with two lens elements to correct for chromatic aberrations. This type of scope is not well suited for astrophotography.

Afocal - A means of taking an image through an eyepiece of a telescope without removing the lens from the camera.

Alt/Az - Altitude Azimuth, a type of telescope mount that moves up and down, left and right as opposed to the smooth rolling motion of an EQ mount which accurately tracks the motion of the stars around the earth.

Amp glow – Amp glow is the glow that some cameras show on a long exposure image. This usually manifests itself in the corners of the image first and then can spread towards the center. A moderate amount of this can be removed using dark frames. Severe cases cannot be corrected.

Aperture - In telescopes, the diameter of the opening at the front of a telescope, usually measured in millimeters. Can also be measured in inches for larger scopes. In camera lenses there is a diaphragm inside the lens that controls the aperture which is sometimes referred to as an F-Stop.

Apochromatic (APO) – A type of refractor extremely well adjusted to remove most or all chromatic aberrations which makes it excellent for astrophotography uses. Can have two, three, or more lens elements. Higher end versions almost always have three or more elements.

Arc Minute – There are 360 degrees in the sky as it goes 360 degrees around us. One arc minute is $1/60^{th}$ of a degree.

Arc Second – Is equal to $1/60^{th}$ of an arc minute.

Artifacts - Errors or unwanted signals in the image.

ASCOM - abbreviation for AStronomy Common Object Model and is a standard in the astronomy equipment industry for control interface design of astronomical equipment such as mounts, focusers, motorized domes, etc.

Astrograph - A type of Newtonian telescope that is designed specifically for astrophotography.

Astrometry – Extremely precise measuring of objects like comets and asteroids.

Astrophotography - Photography of objects in the sky.

Autoguider - A camera and associated equipment used to increase the accuracy of the mount in tracking the stars.

Audio Video Interleave (AVI) – A wrapper for computer video files, can contain a variety of different formats, typically video for Windows formats, and has a file extension of .AVI.

Back Focus – The necessary distance needed to be able to attach a camera onto a telescope focuser, and be able to bring the image projected onto that camera's sensor into focus.

Backlash – Unwanted spacing between gear assemblies usually resulting in some "play" or "slop" with the device. This is normally used to describe issues with a mount but can be applied to anything with gears.

Baffles – Ridges running around the inside of the light path in a telescope to prevent the scatter of light inside the telescope and provide an image with greater contrast.

Bahtinov mask - A mask or cover that goes in front of a telescope with a specific pattern of slits designed to provide easy focusing of point light sources such as stars.

Barlow - An optical device that increases the magnification or reduces the field of view, depending on how you look at it. This trades some image quality and light for more magnification. These plug into the optical train just before the eyepiece.

Bayer matrix - In color one shot cameras (any camera that produces a single color image in one exposure) the pixels are grouped in groups of four, one red, one blue and two green. These are combined to generate the color information for that area of the image. The Bayer matrix is the array of colored filters over the pixels that accomplishes this.

BFA – Bayer Filter Array, see Bayer matrix above.

Bias frame - An image taken with the highest shutter speed possible on a given camera at the same ISO and temperature of the light frames. This is used to subtract the camera's electrical signal present in every frame it takes from the final image.

Binning – A process of combining multiple pixels in order to boost sensor sensitivity at the expense of resolution. For example, 1x1 binning means each pixel counts as one pixel and is in effect not binned, 2x2 binning would take a square of 4 pixels and combine them into one "super pixel".

Binos - Short for binoculars.

Bino-Viewer - A device that allows attaching two eyepieces to a standard telescope so you may view objects in stereo.

Bit - A single bit can be either on or off, representing either 0 or 1. Computers use this as the basic language of everything they do.

Bit depth - This describes a measurement of something like the number of colors an image can contain and is base two mathematics. An example is a 1 bit scale will contain two possible combinations, a 2 bit scale will contain 4, a 4 bit scale will contain 16 and an 8 bit scale will contain 256 bits.

Black point - An area of an image that represents absolute black.

Blooming - In a camera, once a pixel has received as much light as it can handle, the voltage can spill over into adjacent pixels causing them to be brighter than they should.

Bortle scale – Astronomer John Bortle developed a scale of nine levels which represents the "true darkness" of a site, or the amount of light pollution present.

Bulb exposure – A bulb exposure is an exposure where as long as the shutter button is held the camera continues the exposure. DSLRs and other cameras can be used in this mode.

CCD - Short for Charged-Coupled Device, a type of sensor used in digital cameras. In astrophotography it is usually used as a reference to a camera designed and used specifically for astrophotography as opposed to a digital SLR or other multi use digital camera.

Celestial equator - An imaginary line which is basically the equator of the earth projected up into the sky.

Center mark – A dot placed exactly in the center of the primary mirror of a Newtonian to aid in collimation.

Chromatic aberration – Chromatic aberration is the "glowing" or "fringing" of light around bright objects in a telescope. This is caused when light passes through the optical path it is split into its component colors and then rejoined imperfectly at the focal point.

Clip - Clipping an image means you have cut off one end or the other of the image's ability to record data (as can be shown in a histogram). Clipping the highlights for example means that area of the image is pure white and cannot contain any detail. Clipping the darks means that part of the image is pure black and contains no detail.

CMOS - Complimentary Metal Oxide Semiconductor. In astrophotography, a type of sensor in a camera.

Collimation - The act of aligning the optical components of a telescope to make sure all parts of an image combine correctly into one sharp image.

Coma - An optical defect normally present in reflector telescopes that can cause point light sources such as stars to appear to be out of round, presenting like they have the tail of a comet.

Coma corrector - An optical device for reflector telescopes to correct for coma aberrations.

Convolution – A mathematical method of multiplying arrays of numbers to get a third array of numbers. Used in image processing to stretch or resize images.

Corrector plate – The lens on the front of an SCT type telescope that corrects for the spherical aberration created by the spherical mirrors used in that design.

Counterweight – A weight, usually on an equatorial mount, used to balance the weight of the telescope and associated hardware.

Crayford focuser – A telescope focuser that uses smooth bearings and rollers as opposed to gears used in rack and pinion style. They usually come in dual speed (coarse and fine adjustments) and can have adjustable tension.

CRW/CR2 - Canon's RAW image format.

Dark frame - An image taken at the same ISO, shutter speed and temperature as the light frames but with the lens cap/scope cap on, or the shutter closed. This is used to detect the thermal signature of the camera's sensor at these settings so they can be subtracted from your final image.

Dead pixel - Opposite of a hot pixel, a pixel that is stuck in the off position and registers as black regardless of the amount of light applied.

Declination (DEC) – Celestial coordinate measured from the celestial equator north and south of that line, from +90 degrees to the north to -90 degrees to the south, zero being the celestial equator.

Deconvolution - A method of image enhancement that corrects for the bad effects of convolution. This can substantially increase fine details in an image.

Dew heater - Usually a strip that heats up and is wrapped around a telescope near the optics. This warms the optics and prevents dew from forming.

Dew shield - A device attached to the end of a telescope and is like a hollow extension of the telescope tube. This delays the objective from collecting dew, and reduces the intake of extraneous light sources.

Diagonal - A device that has a mirror inside and reflects the image at a 45 degree or 90 degree angle for easier viewing. One side goes into the focuser, the other end holds an eyepiece.

Diffraction - As light passes through a telescope it passes through openings. As light gets near the edges of these openings it is diffracted. This causes stars to appear larger than they actually should.

Diffraction limited – Term used primarily by telescope manufacturers that says that the telescope should perform so that any defect seen will be with the physical characteristics of light and not optical problems with the telescope.

Dispersion – Cause of chromatic aberrations. Prism effect, when light is spread out into its spectrum from white light.

Dobsonian - a type of telescope mount, but usually used as a reference to the entire telescope assembly. These are usually larger Newtonians mounted onto a base that sits on the ground and moves as an alt/az. Like regular Newtonians these are not well suited to astrophotography due to not having enough backfocus.

Doublet – A refractor telescope with two objective lenses.

Dovetail - A metal rail that attaches to the bottom of the telescope, usually by rings that clamp into the telescope tubes or bolts into the bottom of the telescope, which can then be quickly and easily attached to the mount's clamp. Popular dovetail types include Vixen and Losmandy.

DSLR - Digital Single Lens Reflex camera. A type of camera where the user actually looks at the same image that will be recorded on the sensor by means of a mirror and prism that reflects the light from the lens through an eyepiece. When the shutter is opened to take the picture the mirror swings out of the way, the eyepiece goes black as it is no longer receiving the reflected image, and the sensor is exposed.

DSS – Short for Deep Sky Stacker, very popular free program generally used by beginning astrophotographers for stacking images.

Dynamic range - The range from brightest to darkest that a camera can record.

ED – Extra low Dispersion, optical glass corrected for chromatic aberration.

EQ/Equatorial Mount - A type of mount specifically designed to track the stars as they travel around the earth compensating perfectly for their arc in the sky.

Ephemeris – Detailed positional information about planets, their moons, comets and asteroids.

Eyepiece - An optical device that focuses the light exiting a telescope tube in such a way that you can view it with your eye. These typically contain many lens elements in a round cylinder that is inserted into the focuser. The eyepiece can be made to magnify or reduce the image size.

Eyepiece projection - A method of taking a photograph through the eyepiece of a telescope without a lens on your camera. This uses a specific adapter. This can come in handy on telescopes that cannot reach focus using a prime focus adapter.

F-Stop - When using a camera with its lens installed, the aperture is adjustable and is commonly referred to as the F-Stop.

Field flattener - An optical device used primarily on refractors to make sure that the image arrives at the camera sensor perfectly flat. This prevents elliptical images of stars in the corners of the images while the stars in the center may be perfectly round.

Field of view - Commonly represented as FOV. The area of the sky that you can see at one time. Longer focal lengths (more magnification) generally show smaller areas of the sky and hence a smaller field of view. Eyepieces with smaller numbers cause the same effect.

Field rotation – The effect of the image being blurred from the rotation of the sky. This can happen when you use an Alt/Az mount to take long exposures since the Alt/Az mount does not rotate the camera like an EQ mount does.

Filter – A filter is a piece of glass (or Mylar in some solar filters) that alters the light coming through the telescope before the eyepiece or camera. A filter is used for removing light pollution, enhancing certain colors, shooting color images with a monochrome camera and many other tasks.

Finder - A small telescope or other pointing device that helps you quickly orient your telescope towards a particular target. Similar to a gun sight.

Firmware - The software a device uses to tell it what to do. For example, your GoTo telescope software in the hand controller is called its firmware and can be updated on many devices.

FITS format - A file format designated by .FIT (such as .TIF, .GIF or .JPG) specifically designed for scientific purposes. Like RAW or TIF files this stores raw data that does not degrade from repeated editing as do formats such as .GIF or .JPG.

Flats/Flat frame - An image taken with even illumination over the front of the telescope and exposed to present a neutral gray image. This must be taken with the exact same setup as your light frames (same focus setting, same filters, etc) and is used to remove vignetting.

Focal length - The length of a line following where the light travels through a telescope, this is important for calculating parameters such as the FOV and magnification.

Focal plane – An inferred plane at the point where the image from the telescope comes to focus. A camera's sensor is mounted so that it is at the focal plane.

Focal ratio (FR) – The focal length divided by the aperture of the primary objective of the telescope.

Focal reducer - An optical device which reduces the effective focal length and increases the field of view of a telescope, seemingly reducing the magnification. This is usually mounted into the focuser before any eyepieces or cameras.

Focuser - A piece of equipment mounted on the telescope where the light exits. Eyepieces, diagonals, barlows and cameras are mounted into the focuser. Its job is to move the eyepiece/camera/etc back and forth until the light comes into focus at a specific point (your eye or the camera sensor).

FOV – See field of view.

Frames per second (FPS) – The number of image frames captured per second by the device, used in video capture devices.

Full well capacity - A measurement of the total amount of light a photosite can store before saturation occurs.

FWHM – Full Width Half Maximum. The measurement of the angular apparent size of a star, usually used to get the size as small as possible in an image which represents the best possible focus.

Gain - This is a multiplication of the incoming signal. For example, if one photon enters a camera and hits the sensor, setting the gain to 2x will cause the digital signal sent from the camera sensor to say that two photons hit the sensor. Increasing the ISO of a digital camera is increasing the gain.

German equatorial mount (GEM) – Another name for the equatorial mount.

GoTo – A telescope that when properly aligned can point to a celestial object automatically when selected from a catalog or menu.

GPS – Global positioning system, a device or feature used to determine your exact location on the planet.

Grayscale - An image recorded in black, white, and variations of gray with no color information.

Guiding - The act of following a star or other object using either manual corrections (as was the case back before GoTo and tracking mounts) or automatically using guiding equipment such as an autoguider.

Hand controller (HC) – The handheld device used to control your telescope's mount.

HDR - High Dynamic Range. You can use different exposures on different images and sandwich them together to show an image that has too much dynamic range to be captured in one single exposure. M42 is a prime example of a target that needs HDR processing: if you expose correctly for the faint dust lanes on the outer areas, the central core is blown out or clipped; if you expose for the central core, the outer dust lanes are clipped into blackness and can not be seen.

Highlights - Areas of maximum brightness in an image.

Histogram - A graph that shows how an image is exposed. In a normal grayscale histogram the left side is absolute black, the right side is absolute white and there is usually a hump in the graph display somewhere near the center showing the exposure of that image. Color works the same way but shows the intensity of the red, blue and green color channels.

Hot pixel - Opposite of a dead pixel. A pixel that shows exposure information even when shot in complete darkness.

Illuminated reticle eyepiece – An eyepiece with an illuminated crosshair or other centering marker used for precise centering of targets in the field of view.

ISO - International Standards Organization, used to measure the "speed" of film, or the sensitivity of a sensor in a digital camera. As ISO increases, less light is required to "expose" for a given image. This also reduces the signal to noise ratio, increases noise, and reduces the bit depth possible in the image.

JPG/JPEG - Joint Photographic Experts Group. A file format denoted by .JPG (such as .TIF or .GIF) that is very common in digital cameras. Using this format should be avoided because it uses a lossy compression format to reduce file size. This results in huge losses of information and makes it virtually impossible to process well for astronomical uses.

Light frame - A standard picture. Every regular picture you have taken with a regular camera of birthdays, friends and family are all what we call light frames. These are the frames you work with that contain your image data.

Light pollution – Stray light from street lights, signs, windows etc that shine or are reflected up into the air. This is scattered by contaminates and humidity in the air and create a glow effect around cities making it difficult to see outside the atmosphere.

Light year – The distance light travels in a year through a vacuum, approximately 5.87 trillion miles.

Limiting magnitude – The measurement of the dimmest star you can see at zenith which takes into consideration all parameters such as light pollution, weather conditions and optical devices used (if any).

Lossless compression - Certain file formats such as PSD and TIF employ compression methods that preserve 100% of the data while decreasing the file size.

Lossy compression - Formats such as .GIF and .JPG use lossy compression which throws away data that it does not think is needed to display the image.

LRGB - When shooting a monochrome camera and creating a color image you need to shoot at least one image with a red filter, one image with a green filter and one image with a blue filter. These are combined together into one color image. The L in LRGB stands for luminance and is used to increase detail in an image. The Luminance frame is the detail frame and can be shot in very high resolution. The color can be shot at lower resolutions and combined with the luminance to create a high resolution color image. You can use this idea to increase your ability to stretch images as well.

Luminance – The recording of brightness or intensity of light. Typically this is the high resolution/detailed portion of an image.

Magnitude - A measurement of the brightness of an object. An increase in one magnitude is approximately 2.5 times as bright. The lower the number on the scale, the higher the magnitude.

Maksutov Cassegrain telescope – See MCT below.

Maksutov Newtonian – Similar to a Maksutov Cassegrain except they are designed as a Newtonian configuration with the focuser near the front of the scope.

MCT - Maksutov Cassegrain Telescope, a type of telescope that has a sealed front end which is actually a corrector lens called a meniscus, two mirrors and has its eyepiece in the rear.

Megapixel - Roughly one million pixels.

Meridian - An imaginary line dividing the west and east halves of the sky running from the north celestial pole directly overhead to the south celestial pole.

Meridian Flip - Meridian Flip is the act of re-orienting the scope on an EQ mount so it can continue to track past the meridian. This "flips" the scope around to pointing the other direction at roughly the same spot on the meridian. Going past the meridian without flipping can cause the scope to run into the mount, cables to come loose, and many other really bad things.

Micron – One millionth of a meter or 0.001mm.

Mirror cell – The frame that holds the primary mirror assembly.

Mirror lock(DSLR) – Some cameras have the ability to lock the mirror in the up position to minimize camera vibration when the shutter is tripped. This can be very useful shooting brighter objects like the moon but is ignored in long exposure work as the amount of time the camera is vibrating due to the mirror slamming open is miniscule compared to the overall exposure time.

Mirror lock(SCT) – Some SCT type telescopes have the ability to lock the mirror once the image is in focus to prevent the mirror from "flopping" or moving as the orientation of the telescope changes.

Monochrome – Technically means one color, meaning either black or white. "Monochrome" cameras are actually grayscale in that they produce black, white and many different shades of gray.

Mosaic - The act of shooting multiple images in a grid pattern and stitching them together to allow you to shoot a larger field of view than you could normally.

Mount - The mount is the geared (and sometimes motorized) device that is typically attached to the top of a tripod and then has the telescope attached to it. It is the mount that allows you to point the telescope at different objects without moving the tripod, and (when motorized) tracks objects across the sky.

Narrowband - Using special filters you can capture the emissions from certain gasses such as hydrogen alpha, sulpher and oxygen. These can be used much like LRGB imaging to create faux color images of high resolution. This method can also overcome all but the worst light pollution situations and can even allow you to shoot on nights with a full moon to some degree.

Near Earth Object (NEO) – An object such as a comet or asteroid which will pass in close proximity to earth.

Newtonian - A type of reflector telescope that has two mirrors in a hollow tube. The front of the telescope is open to the elements and the back is sealed. The eyepiece is near the front of the scope. These are usually not suitable for astrophotography unless they are designed as an "astrograph" as they will not bring a camera to focus without modifications or the use of a Barlow.

North celestial pole (NCP) – The point in space very close to Polaris where a line drawn from the exact southern to northern poles would extend into space with the earth revolving around that line.

Nyquist theory - States that when converting frequencies, the sampling rate should be 2x the highest frequency to get an accurate conversion and preserve all the data.

Objective lens – Also called the primary objective, the large front lens of a refractor telescope.

Off axis guider (OAG) – A method of mounting a guide camera so that it shares most of the same optical path as the imager, picking off a small amount of light usually from a mirror mounted in the light path.

One shot color (OSC) - Any camera that creates a color image from a single exposure.

Opposition – Opposition is when a planet is closest to the earth and is directly on the other side of earth from the sun.

Optical train - Anything that is directly in the path of light from the stars to your eye or camera sensor is considered "in the optical train". Could be called the optical path as well.

Optical tube assembly (OTA) – Also referred to as the OTA, this is the main tube of the telescope not including any mount, pedestal, pier or tripod.

Parfocal – Applies to both eyepieces and filters and means that if you exchange one filter (or eyepiece) for another, you will remain in nearly perfect focus. Not all filter sets or eyepiece sets are parfocal.

Periodic error (PE) - Errors in the manufacturing process of the gears and drive assembly in an EQ telescope mount results in repeating errors in the tracking of the mount. These can be removed with software that contains PEC code.

PEC - Periodic Error Correction. Software that corrects for periodic error.

Photometry – The measurement of apparent magnitude of objects such as comets, asteroids and stars.

Photon - For the purposes of discussion in this book, a photon is a single particle of light.

Photosite - The technical name for the tiny part of the sensor in a digital camera sensor that when exposed to light records a signal. Typically called a pixel.

Piggyback - Mounting a camera with a lens on a telescope in such a way as it is not shooting through the telescope but is instead just using it as a tracking mount.

Pixel - A single dot in an image.

Pixel size - The physical size of a photosite on the sensor of a camera, measured in microns.

Plate solve – Refers to Plate Solution, or finding the absolute position and motion of an object. Some applications such as TheSkyX Professional offer a plate solve feature where it can look at your image and tell you exactly what is in the frame.

Point light source - Stars are considered point light sources because regardless of their magnification they are so far away they will always appear as a single point of light.

Polar alignment – Aligning the "polar axis" of an equatorial mount to either the northern or southern celestial pole so that the mount can track celestial objects precisely.

Polar scope - A small telescope usually built into the mount which allows for precise pointing of the mount's right ascension axis to the north or south celestial pole.

Prime focus - Attaching a camera without a lens in such a way that the image from the telescope is directly projected onto the sensor of the camera.

Quantum efficiency (QE) - A measurement of the percentage of photons which hit a photosite versus how many are detected.

Rack and pinion focuser – A less expensive and typically less accurate style of focuser.

RAW - A RAW file is a file that contains the relatively unaltered, unmodified data directly from the camera's sensor.

Rayleigh scattering – The scattering of different wavelengths of light by the molecules in the atmosphere. This scattering is the reason the sky appears blue.

Resolving power – 4.56/(inches of aperture of the telescope)=resolving power of the telescope in arc-sec. Note that this does not take into consideration obstructions such as secondary mirrors.

Reticle - Crosshairs or other markings that allow you to precisely center a target in your field of view. Sometimes included inside eyepieces and finder scopes.

Red dot finder - A type of finder that uses an illuminated red dot as a reticle.

Refractor - A type of telescope that has an objective lens on the front end and an eyepiece or camera at the other. Light passes straight through without being reflected unless a diagonal is used.

RGB - Red, Green, Blue. One shot color cameras shoot everything as a combination of these three primary colors. When shooting monochrome images and wanting to end up with a color image, you shoot at least one frame with a red filter, one with a green, and one with a blue and then combine them to create a full color image.

Right ascension (RA) – Celestial coordinate measured from west to east in hours, minutes and seconds. As the earth turns each hour, 15 degrees of arc pass.

Saturation – The point at which you cannot record any more data. This may refer to the full well capacity of a CCD camera or the maximum value a pixel can store.

Schmidt Cassegrain Telescope (SCT) - a type of reflector that has a sealed front, two mirrors and has its eyepiece in the rear of the scope.

Seeing - A measurement of the conditions of the atmosphere as it relates to being able to view or image an astronomical object. An easy method to determine the seeing conditions is to look for stars twinkling; the more they twinkle, the worse the seeing.

Sidereal rate – 23 hours, 56 minutes and 4 seconds is one sidereal day which is why the stars are never at the exact same place at the exact same time every night and seem to "advance" across the night sky every night all year long. This is the rate at which your telescope must track to remain aligned with your target.

Signal to noise ratio (SNR) - The ratio of signal (what you are trying to capture in the image) to noise (electrical signals inherent to the camera generating the image). The higher the SNR, the easier it is to stretch an image and bring out the detail of your target.

Slew – The process of your telescope moving to and from targets.

South celestial pole (SCP) – The point in space very close to Sigma Octantis where a line drawn from the exact northern to southern poles would extend into space with the earth revolving around that line.

Spider vanes - Small strips of metal or plastic in the front of a Newtonian telescope which supports the secondary mirror in the optical path.

Stacking - Taking several images and combining them in such a way as to increase the signal that you want to keep while reducing the noise levels that you do not.

Strehl ratio – Gives a ratio as compared to a theoretically perfect optical system. For example, a Strehl ratio of .90 is 90% as good as a theoretically perfect optical system.

Stretching - Taking an image and manipulating the data so that details that were too dark to see are now light enough to be visible through compression of the grayscale or color scale.

T-Ring – An adapter that mates with a removable lens camera on one side and has threads on the other side to attach to the telescope or other device.

Thermo Electric Cooler (TEC) – Electric cooling device used with some CCD and DSLR cameras.

TIF - A file type (like .GIF and .JPG) to store image files. TIFs are excellent because they are lossless formats. They are however far larger than JPG or GIFs.

Tracking - The ability to follow an object as it appears to travel across the sky.

TSX - Abbreviation for TheSkyX, a planetarium, telescope control and planning application for amateur and professional use from Software Bisque Inc.

United States Naval Observatory (USNO) – The standard for timekeeping in the United States.

Vignetting - The effect of the edges of an image being darker than the center due to obstructions or optical imperfections.

Well depth - A measurement of the total amount of light a photosite can store before saturation occurs.

White point - A part of an image that represents pure white.

Zenith – The point directly overhead.

5.4: Other books by the author

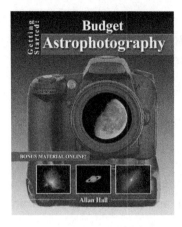

Want to take a few snapshots of the beautiful objects you are viewing without spending a small fortune? Already have a camera but you can't seem to get a good image and want to know why?

This book will answer those and many other questions while giving you a quick and reasonably easy introduction to budget astrophotography. In addition, save more money by seeing how to make a lot of items you may find useful.

http://www.allans-stuff.com/bap/

If you decide that you want more than quick snapshots, you want big beautiful prints to hang on your wall, this is the book for you.

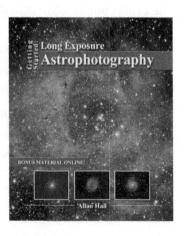

From required and optional equipment, through the capture process and into the software processing needed to create outstanding images, this book will walk you through it all.

http://www.allans-stuff.com/leap/

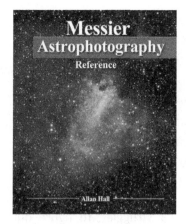

You decide that you want to take images of celestial targets, but need a little help with the targets? This book discusses all 110 Messier targets and includes descriptions, realistic images of each target, star charts and shoot notes to help you image all 110 of the objects yourself.

http://www.allans-stuff.com/mar/

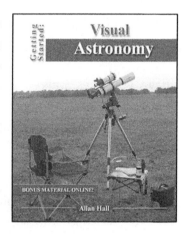

If you have ever wanted to view the wondrous objects of our solar system and beyond, here is the how-to manual to get you well on your way. From purchasing your first telescope, through setting it up and finding objects, to viewing your first galaxy, this book contains everything you need. Learn how to read star maps and navigate the celestial sphere and much more with plenty of pictures, diagrams and charts to make it easy. Written specifically for the novice and assuming the reader has no knowledge of astronomy makes sure that all topics are explained thoroughly from the ground up. Use this book to embark on a fantastic new hobby and learn about the universe at the same time!

http://www.allans-stuff.com/va/

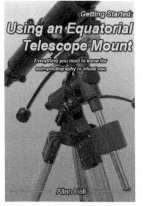

Many midrange and high end telescopes come on equatorial mounts. These mounts are fantastic for tracking celestial objects. Someone who wanted to take pictures of objects in the night sky might even say they are required for all but the most basic astrophotography. The problem is that they can also be unintuitive and require some knowledge to use.

If you have ever struggled to figure out how to use an equatorial telescope mount, this is the book you always wished you had.

http://www.allans-stuff.com/eq/

Have you ever wanted to take a picture of a solar or lunar eclipse but didn't know where to start? This is the book for you!

Whether you want to snap a few pictures with your smartphone or have a telescope with solar filters, this book will help you prepare, find the perfect location, and get those incredibly rare and beautiful images.

http://www.allans-stuff.com/eclipse/

The Dobsonian telescope is one of the most popular styles of telescopes for beginner to intermediate amateur astronomers out there, with good reason. These telescopes provide excellent views for a modest investment, and are also very easy to setup and use.

This book will make sure that before you make your investment you know which telescope meets both your needs, and budget. In addition, you will feel comfortable not only purchasing one, but using one as well.

http://www.allans-stuff.com/dob

Once you know you want to pursue astrophotography, how do you know which of the tens of thousands of possible objects your beginner equipment can take successful images of? How about which ones are the right size for your equipment? What are the raw images supposed to look like when they come out of the camera?

Written specifically for the beginning astrophotographer with beginner equipment such as a small refractor, small reflector, or similar telescope on an EQ mount using a DSLR camera, this book will help you start capturing stunning images quickly and easily.

http://www.allans-stuff.com/50best

So you've decided to write a book and get into non-fiction publishing. Now you find yourself faced with the seemingly infinitely harder second step – actually bringing the idea to market. In today's brave new world of self-publishing and open creative markets, it is both an inviting and potentially intimidating arena for authors hoping to turn their non-fiction books into a meaningful source of income. This is a daunting task because it  involves a blend of several disciplines that aren't necessarily part of an author's quiver of arrows. Most crucial among these are marketing and digital publishing, each of which requires fluency in fields that authors may or may not have experience in.

http://www.allans-stuff.com/ck/

WordPress is the perfect tool to help you build the website you've always wanted. But the 'help' aspect which is built into it isn't always the right thing for someone who is just getting started.

What you need, and what this book will provide, is a book that shows you how to get off the ground and then build on that knowledge to give you a secure and usable website.

http://www.allans-stuff.com/wp/

Information Technology is an area which is constantly on the move, sometimes at a speed which is dizzying and difficult to keep pace with. In particular **data recovery** can be one of the more complex problems you might encounter. The sheer amount of information is often overwhelming and confusing.

Data Recovery for Normal People is a new book which aims to make this process a lot simpler. Designed for both beginners who have little knowledge of technical issues and for those who may own their own computing business and want to learn more.

<u>http://www.allans-stuff.com/dr/</u>

How to find objects in the night sky

5.5: NOTES:

NOTES:

How to find objects in the night sky

How to find objects in the night sky

Made in United States
North Haven, CT
13 September 2023

41503526R10065